Quick Start CNC Operation (+)

Basic Shop Math

The intent of this writing is to prepare all learners with Basic Shop Math problems using the Texas Instruments TI-30XA calculator.

By equipping all learners with the same calculator, we can move forward much faster with the simple mathematical calculations needed.

This calculator has all the necessary functions to complete all computations for the simple adding, subtraction and division of decimal as well as fractions plus any trigonometry needed.

Mike Willard   December-2010

Basic Math

Basic Math

Types of Numbers:

Every type of shop math problem you'll solve is based on the understanding of whole numbers. If you're already familiar with some of the material in this unit, you should find that it's a helpful review.

Whole Numbers..Zero and the numbers used for counting: 1, 2, 3, 4, 5.....

Fractions..Numbers that also represent part of a whole, such as 7/16 or 3/4.

Decimals..Numbers that also represent a part of a whole, but are written differently than fractions, such as .4375 or .7500.

Everyday applications of types of numbers

Whole Numbers

- Expressions of temperature: 72 degrees Fahrenheit.
- Sports scores: 28 to 17 for a football game, 76 strokes for 18 holes of golf, 5 to 2 for a baseball game.

Fractions

- Measurements taken with most inch rules: 3/4 in., 5 3/8 in.
- Amounts of ingredients in recipes: 1/3 cup, 1/2 teaspoon.

Decimals

- All expressions of money: $24.95, $.89
- Gallons shown on a gasoline pump: 12.6, 18.24

Can you think of other everyday or work place examples of each type of number?

Basic Math

Digits and Place Values:

Most CNC machines use a three/four place format for inch Cartesian Coordinate values. This means that the 3/4 number system allows three places to the left of the decimal point and four to the right.

Whole numbers can be input from 000. to 999. and fractional numbers from .0000 to .9999 are allowed.

As a CNC programmer, we want to program our part to four decimal places to the mean dimension. That means, if our math is correct, we will be mathematically perfect to four decimal places and tolerance is a non-factor.

To achieve this we will take the numbers from our calculator as given and "truncate" them to four places rather than round up or down.

- An example of this would be: 27.9025, the spoken word would look like this: 27 inches, nine hundred and two thousands and 5 tenths.

1. Write down 31 inches, one hundred twenty five thousands and six tenths. _____

2. Give the spoken word 4.9568 _____
_____

Basic Math

Negative Numbers:

Every positive number has a negative counterpart that represents values on the other side of the Cartesian Coordinate Zero.

- A negative number always has a minus sign.
- A positive value does not need the (+), it is assumed positive.

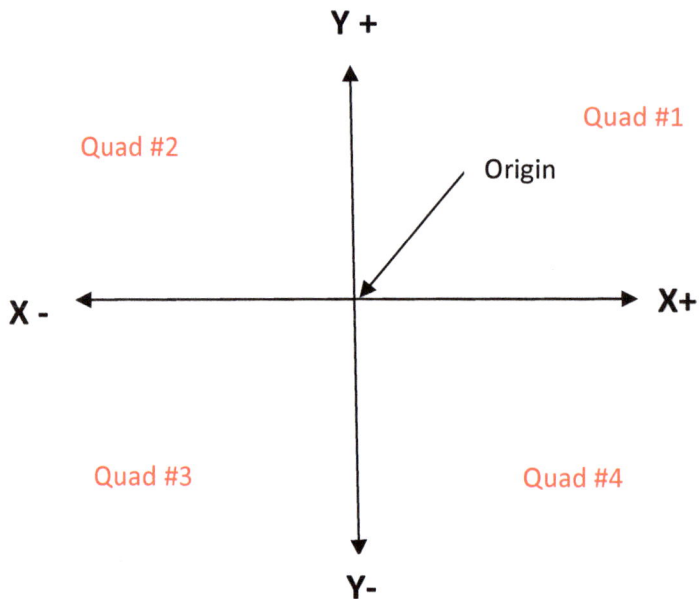

There are four quadrants associated with the Cartesian Coordinate system.

- Quadrant #1   upper right uses +X and +Y values.
- Quadrant #2   upper left uses -X and +Y values.
- Quadrant #3   lower left uses –X and –Y values.
- Quadrant #4   lower right uses +X and –Y values.

Basic Math

Adding Whole Numbers:

Adding and subtracting whole numbers is probably the easiest form of mathematics you will encounter. A calculator is not always needed, however, It would behoove you to use one so as not to make carless mistakes. With all the numbers needed to produce a CNC program, it would be wise to use whatever necessary to eliminate errors.

Problem:

The first two shifts of the day produced 7,369 parts combined, and the third shift produced 2,833 parts. You need to determine the total production for the day.

Solution:

$$7,369 + 2,833 = 10,202$$

Quiz:

3) 47,612 + 3,250 = _____

4) 1,250 + 89 = _____

5) 12,872 + 19,638 = _____

6) 1,626,000 + 458,500 = _____

7) 893 + 107 = _____

8) 40,930 + 627,117 = _____

Basic Math

Subtracting Whole Numbers:

Subtraction does not share the same properties as addition. So, you must set up subtraction problems carefully in order to solve them correctly.

- Subtracting two numbers in a different order will produce a different answer.

$$15 - 3 = 12 \quad \text{but} \quad 3 - 15 = -12$$

- Typical subtraction usually requires that you subtract a smaller number from a larger number.

$$27 - 11 = 16$$

- With CNC and the Cartesian Coordinate system, where you have an "Origin" or program "Zero" that has a minus side as well as a positive side.

Example: If you have a positive 15 in the X axis and want to subtract 19 in the minus direction you would get a negative 4 in the X axis for your position.

$$15 - 19 = -4$$

Quiz:
9) $6,872 - 2,362 =$ _____

10) $8,798 - 8,424 =$ _____

11) $4,687,331 - 4,084,612 =$ _____

12) $5,580 - 117 =$ _____

13) $19,121 - 639 =$ _____

14) $468,000 - 9,775 =$ _____

15) $129,875 - 2,250 =$ _____

16) $587 - 125 =$ _____

Basic Math

17)   The daily output from three production lines is 316 parts 298 parts, and 312 parts. What is their combined output for the day?

18)   A production area needs to complete four orders for parts before the end of the week. The orders are for 1,026 parts, 68 parts, 144 parts, and 5,000 parts. How many total parts must be produced to fill all four orders?

19)   By Wednesday, 4,912 parts had been completed in the production area. How many more parts must be produced to fill the four orders?

20)   The original cost of a shop machine was $9,628. Two years later, the value of the machine had dropped by $3,452. What was the value of the machine two years after it was bought?

Basic Math

Adding & Subtracting Problems:

21)   An electronic gage is preset to display a reading of zero at a part feature's nominal (ideal) value of 162mm, when that feature is measured, the display shows –3. What is the actual dimension of the part measured?

- Solve the following addition and subtraction problems.

22) –286 + 184 = _____   27) 26 + (–48) = _____

23) – 176 – 22 = _____   28) 98 (–104) = _____

24) – 85 + 200 = _____   29) 125 + (–55) = _____

25) 596 + 27 (–37) = _____   30) – 4 + 12 = _____

26) (–31) + 47 = _____   31) 27 + 11 (–30) = _____

Basic Math

Basic Multiplication:

Multiplication is a shortcut for finding the total of two or more equal numbers.

- Which problem do you think would be faster to solve? (They're actually the same problem, expressed differently.)

  $$12 \times 7 \quad \text{or} \quad 12 + 12 + 12 + 12 + 12 + 12 + 12$$

There are several ways to write multiplication problems, but the first is the most common.

$$3 \times 5 = 15 \qquad 3(5) = 15 \qquad (3)5 = 15$$

  o The numbers you're multiplying (3 and 5) are called factors.
  o The answer (15) is called the product.

Two properties of multiplication make it especially easy. (Addition has the same properties.)

  o You can multiply factors in any order and get the same answer.
    $$5 \times 3 = 15 \quad \text{and} \quad 3 \times 5 = 15$$
  o You can group factors in any way and get the same answer.
    $$(3 \times 5) = 15 \times 2 = 30 \quad \text{and} \quad 3 \times (5 \times 2) = 30$$

Basic Math

Basic Multiplication:

Zero and 1 are unique factors.

The product of 0 and any number is always 0. So, the product of any multiplication problem that has 0 as a factor is 0.

$$3 \times 0 = 0 \qquad (3 \times 4) \times (2 \times 0) = 0$$

The product of 1 and any number is always that number.

$$3 \times 1 = 3 \qquad (3 \times 1) \times (1 \times 4) = 3 \times 4 = 12$$

Problem:

A machine on your production floor produces 142 pieces per hour. How many pieces could it produce in three eight-hour shifts (or 24 hours)?

Solution:

$$8 \times 142 = 1,136 \times 3 = 3,408$$

Basic Math

Basic Multiplication:

Problem:

For the same machine that produces 142 pieces per hour, you need to determine the number of pieces it can produce in 80 hours.

Solution:

$$142 \times 80 = 11,360$$

Practice Multiplying Whole Numbers

32. $216 \times 43 =$ _____

33. $721 \times 50 =$ _____

34. $893 \times 107 =$ _____

35. $62 \times 6001 =$ _____

36. $21 \times 136 =$ _____

37. $5 \times 109 =$ _____

38. $1,250 \times 89 =$ _____

39. $6,000 \times 62 =$ _____

40. $9,642 \times 33 =$ _____

41. $621 \times 3 =$ _____

42. $10 \times 528 =$ _____

43. $7 \times 394 =$ _____

Basic Math

Basics of Division:

Division is the reverse of multiplication. So, for every division problem there's a corresponding multiplication problem.

$$18 \div 2 = 9 \qquad\qquad 9 \times 2 = 18$$

- o The number you're dividing, (18) is called the dividend.
- o The number you're dividing by, (2) is called the divisor.
- o The answer is called the quotient, (9). If the divisor doesn't divide into the dividend evenly, the number left over is the remainder.

Division does not share the properties of multiplication. So, you must set up division problems carefully in order to solve them correctly.

- o Changing the order of numbers in a division problem will change the answer.

$$8 \div 2 = 4 \quad \text{but} \quad 2 \div 8 = .25$$

- o Changing the way the numbers are grouped will change the answer.

$$(24 \div 6) = 4 \div 2 = 2 \quad \text{but} \quad 24 \div (6 \div 2) = 8$$

Basic Math

Basics of Division:

Problem:

A machine on your production floor produced a total 0f 4,896 pieces during the last 9 shifts. How many pieces has it produced per shift, on average?

$$4{,}896 \div 9 = 544$$

Solution:

Quotient x Divisor = Dividend

Problem:

You're running an operation that produces 72 pieces per hour, and you need to fill an order for a total of 6,125 pieces. Assuming ideal conditions, with no downtime, how long will it take to complete the run and fill the order?

Solution:

$$6{,}125 \div 72 = 85.0694 \text{ or } 85+ \text{ hours}$$

---

44. $7{,}622 \div 8 =$ _____

45. $6{,}710 \div 11 =$ _____

46. $2{,}292 \div 3 =$ _____

47. $13{,}095 \div 27 =$ _____

48. $14{,}400 \div 120 =$ _____

49. $131 \div 6 =$ _____

50. $975{,}000 \div 225 =$ _____

51. $5{,}138 \div 466 =$ _____

52. $127 \div 9 =$ _____

Basic Math

Order of Operations:

For shop math problems that require several operations, you must follow a certain order to solve them correctly. When parentheses are shown, work whatever operations they contain first. If not, always follow the order of operation:

Multiply   Divide   Add   Subtract

Think of a phrase to help you remember the order, such as:

My Dear Aunt Sally   or   My Dog Acts Strange

Problem:

You need to find the solution to : **25 – 3 x 3 + 12 ÷ 2**

Solution:

25 – 3 x3 + 12 ÷ 2	Multiply 3 x 3
25 – 9 + 12 ÷ 2	Divide 12 ÷ 2
25 – 9 + 6	Add 9 + 6
25 – 15	Subtract 25 – 15

Answer: 10

Knowing the correct order of operation is particularly important when you're using a calculator. If you simply enter a string of numbers and signs, a calculator will do each operation as it's entered, which is often not correct.

Basic Math

Multiplying & Dividing Whole Numbers:

53. Thirty-five pieces of steel, each 24cm long, are needed for the job. What is the total length of steel required?   (excluding loss for width of saw blade)

54. During a four-hour period, one production line ran 52 parts per hour, another production line ran 55 parts per hour. What is the total number of parts produced in the four hours?

55. Components are shipped in crates that each contains 138 units. How many components are there in 24 full crates?

56. After the components are machined, they are repacked into crates that hold 18 components each. How many crates are required for the total number of machined components?

Basic Math

Multiplying & Dividing Whole Numbers:

56. How many pieces of wire, each 14 inches long, can be cut from a 266-inch length of wire?

57. How many parts with an overall length of 8 inches can be produced from a 112-inch length of stock?

58. If it takes approximately 3 minutes to produce one 8-inch part, approximately how much time will it take to use the entire 112-inch length of stock?

59. A job requires 23 pieces of metal tubing that's 17 inches long and 7 pieces of tubing that's 32 inches long. What is the total length of tubing needed for the job?

60. How many parts, 1.25 inches in length can be produced from a 36 inch length of bar stock when a 1/8 in. wide cut-off blade is used and production needs at least 1.0 inch remnant left in the chuck jaws for grip?

Basic Math

Multiplying & Dividing Whole Numbers:

61. If 22 parts can be produced from one bar of stock, how many bars of stock would be needed to run an order of 1,650 parts?

62. A company leases a variety of office and production floor machines for an annual cost of $70,476.00. What are the leasing costs per month?

63. The area of a rectangle is calculated by multiplying the length of one of the sides by the length of one of the longer sides. The answer is expressed as the square of the unit of measure. For instance, the area of a rectangle with sides of 2 inches and 3 inches is: 2 x 3 = 6 square inches. What is the area of a rectangular room that's 18 feet by 21 feet?

64. What is the perimeter of a lot that is 128 feet by 496 feet?

65. What is the area of same lot?

Basic Math

Understanding Fractions:

On the production floor, you may see fractions used in a number of ways. They often identify sizes of drills, stock and standard cutting tools. They're also used for dimensioning part features that do not require tight tolerances.

This unit will help you become familiar with different types of fractions so that you can work with them in typical shop problems.

Definition:
Fraction – A way to express one or more equal parts of a whole.

Example: 3 / 8

The upper number (3) is the Numerator or number of equal parts you're working with.

The lower number (8) is the Denominator or the number of equal parts in the whole.

- When a fraction has the same numerator and denominator, it equals one.

$$2 \div 2 = 1 \quad 4 \div 4 = 1 \quad 16 \div 16 = 1$$

- On production floors, fractions are most often used for dimensions in inches, and have denominators of:

2 4 8 16 32 64 or 1/2, 3/4, 5/8, 1/16, 3/32, & 5/64

Think of the markings on rulers and measuring tapes you use at home. Typically, each inch is divided into fourths, eights, and sixteenths. Some may also have marks for thirty-seconds of an inch.

Basic Math

- o **Proper Fractions** – The numerator is smaller than the denominator.

    3 / 8   1 / 16   15 / 32

- o **Mixed Number** – A combination of a whole number and a fraction.

    1 1/8   5 3/16   9 17/16

- o **Improper Fraction** – The numerator is equal to or greater than the denominator.

    4/4   11/8   19/16

Practice: Changing Fractions to Decimals

67. 55/16 = _____   73. 12 3/4 = _____   79. 37/64 = _____

68. 10 1/2 = _____   74. 25 1/4 = _____   80. 291/32 = _____

69. 6 3/8 = _____   75. 91/16 = _____   81. 4 5/8 = _____

70. 18/24 = _____   76. 24/50 = _____   82. 22/64 = _____

71. 28/32 = _____   77. 8/64 = _____   83. 3/32 = _____

72. 14/64 = _____   78. 6/32 = _____   84. 35/50 = _____

Basic Math

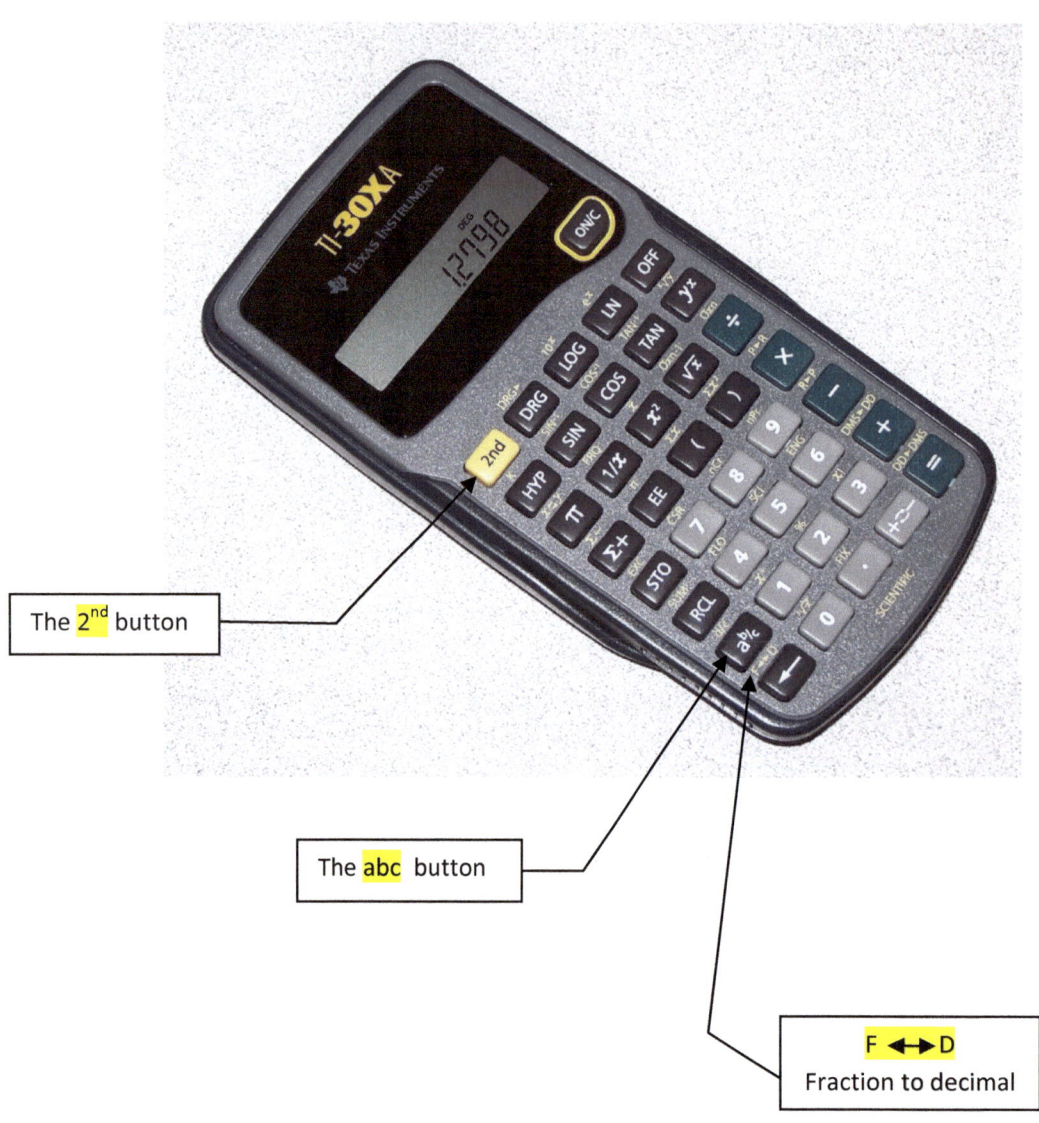

The 2nd button

The abc button

F ←→ D
Fraction to decimal

Basic Math

Adding and Subtracting Fractions (using the TI-30)

If you are working with parts that are dimensioned in fractions, you'll need to add and subtract these numbers. In many cases, these operations are nearly as simple as adding and subtracting whole numbers.

With the calculator, you can add and subtract and retain fractions, but most likely you will have to change them over to decimal dimensions.

Example: 21/32 + 15/32 = 1 1/8 or 1.125

This is done on the calculator by using the abc button in place of the (/).

Example: 21 abc 32 + 15 abc 32 = 1 1/8

To change 1 1/8 fraction to decimal: press the 2nd button on the calculator then the fraction to decimal button in yellow.

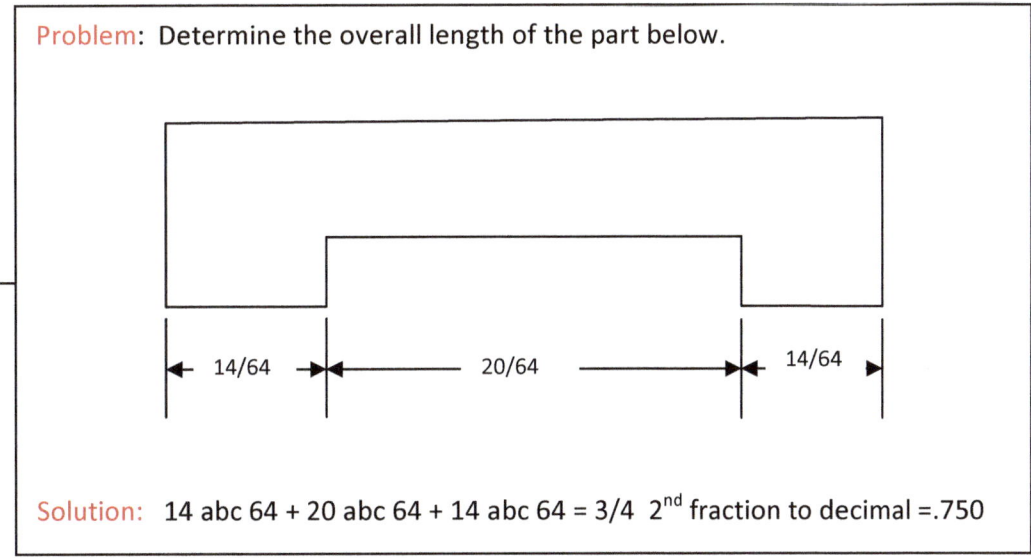

Problem: Determine the overall length of the part below.

14/64   20/64   14/64

Solution: 14 abc 64 + 20 abc 64 + 14 abc 64 = 3/4 2nd fraction to decimal =.750

Basic Math

Adding and Subtracting Fractions (using the TI-30)

Problem:

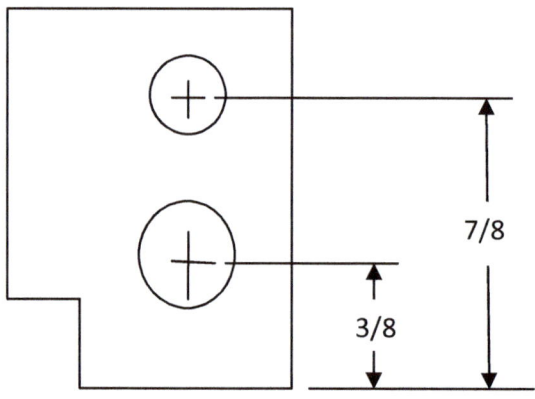

You need to determine the distance between the centerlines of the holes on the part drawing above.

Solution: 7 abc 8 – 3 abc 8 = 1/2   then 2nd and fraction to decimal. = .5

Practice: (give decimal value)

84. 9/16 + 3/32 = _____

85. 9/32 – 3/16 = _____

86. 25/64 – 12/32 = _____

87. 3/4 + 5/8 = _____

88. 11/32 + 15/64 = _____

89. 17/32 – 3/16 = _____

90. 7/16 – 5/32 = _____

91. 3/4 + 5/16 + 3/32 = _____

92. 5/16 + 3/4 + 7/8 = _____

Basic Math

Adding and Subtracting Fractions (using the TI-30)

Problem: Determine the overall length of the part shown below.

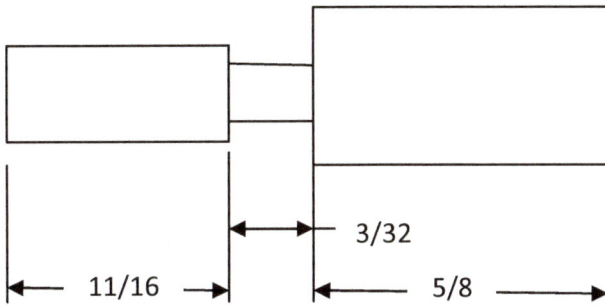

By using your calculator, you do not have to find the least common denominator.

Solution: 11 abc 16 + 3 abc 32 + 5 abc + 8 = 1 13/32

Then 2nd and fraction to decimal = 1.4062

Practice: (decimal values)

93. 9/16 + 3/8 + 3/4 = _____   96. 7/8 – 11/16 = _____   99. 1/2 + 7/8 + 3/4 = _____

94. 5/32 + 3/16 + 1/4 = _____   97. 39/64 – 15/32 = _____   100. 7/8 – 25/32 = _____

95. 3/8 + 3/16 + 3/32 = _____   98. 9/64 – 1/8 = _____   101. 1/2 + 1/4 + 7/8 = _____

Basic Math

Adding and Subtracting Fractions (using the TI-30)

Problem: Mixed Numbers

102. 2 3/8 + 4 1/4 = _____   105. 4 1/4 + 3 7/8 = _____   108. 12 3/16 + 6 7/8 + 5/32 = _____

103. 38 9/16 – 11 1/2 = _____106. 9 1/2 + 28 7/16 = _____109. 1/8 + 3/32 + 3/4 = _____

104. 16 1/2 – 2 13/16 = _____107. 2 9/32 – 3/4 = _____   110. 9/64 + 4 3/8 + 2 5/16 = _____

Find distances X and Y on this drawing.

X = _____

Y = _____

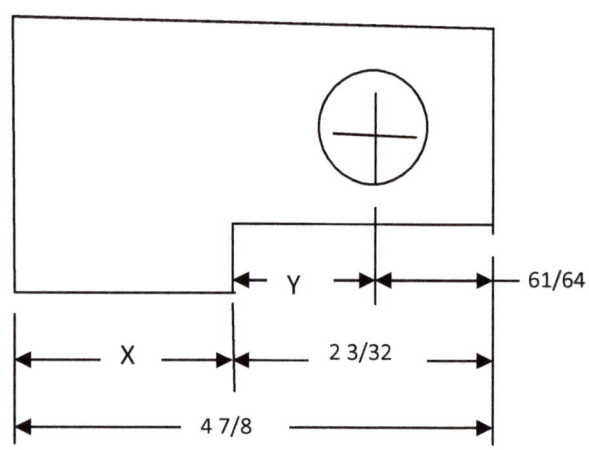

Solve: (decimal values)   111. 10 1/4 + 3/16 + 2 7/8 = _____   112. 5/32 + 39/64 – 11/16 = _____

113. 12 7/16 – 7 31/32 = _____

114. You're producing three parts with overall lengths of 4 5/16, 2 7/8 and 5 5/32 from a 15 inch length of stock. How much stock will you use? _____

115. What length of stock will be left over? _____

Basic Math

Multiplying and Dividing Fractions: (using the TI-30)

Although you probably won't multiply or divide fractions often, you should be able to solve problems that require these operations. As you'll see, multiplying or dividing fractions is simpler than you may think.

Practice examples: (decimal answers)

116. 3/8 x 7 = _____

117. 5/16 x 20 = _____

118. 10 1/2 x 38 = _____

119. 1 1/2 x 15 = _____

120. 4 3/4 x 6 = _____

121. 9/16 x 12 = _____

122. 15/32 x 3 = _____

123. 2 1/2 x 7/8 = _____

124. 18 x 3/64 = _____

Problem:

You have a stack of aluminum sheets that's 2 7/8 inches tall. Each sheet is 1/16 inch thick. How many sheets are in the stack?

Solution: 2 abc 7 abc 8 ÷ 1 abc 16 = 46

---

125. 15 ÷ 3/4 = _____

126. 36 ÷ 9/16 = _____

127. 33 ÷ 11/32 = _____

128. 27 1/2 ÷ 4 = _____

129. 90 ÷ 45/64 = _____

130. 12 ÷ 15/64 = _____

131. 120 ÷ 5/8 = _____

132. 10 1/4 ÷ 8 = _____

133. 46 7/8 ÷ 3 = _____

Basic Math

Multiplying and Dividing Fractions: (using the TI-30)

134. A shipping crate is designed to hold a maximum of 75 lb. How many parts can it hold if each part weighs 6 $\frac{1}{4}$ lb.?

135. The capacity of a diesel fuel tank is 22 gallons. When reading shows that the tank is 3/8 full, how many gallons of fuel does it contain?

136. If one tie rod weighs 1 5/16 lb., how many rods are there in a quantity of rods that weigh 42 lb.?

137. How many 5 9/16 inch long parts can be cut from a plastic extrusion that's 267 inches long?

Basic Math

Multiplying and Dividing Fractions: (using the TI-30)

Find distances X and Y on the drawing below. (all dimensions shown are in inches.)

X = _____   Y = _____

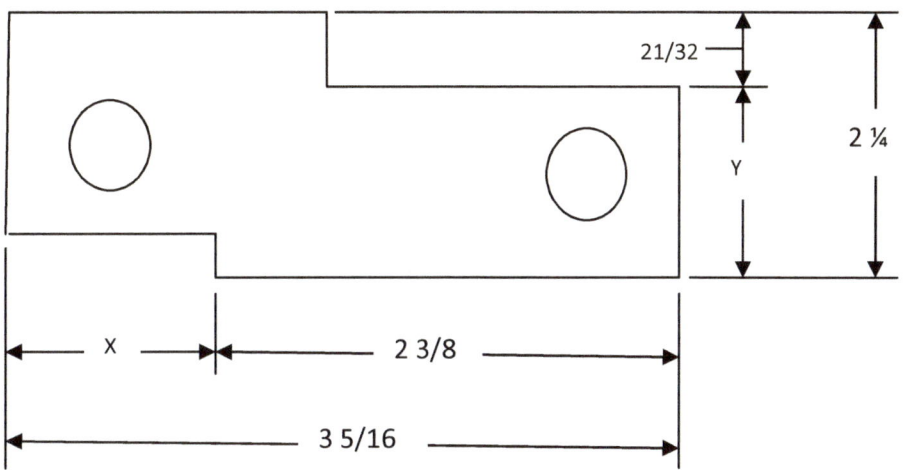

       The perimeter of a rectangle, or the distance around its edges, can be found with the formula (2 x length of one of shorter side) + (2 x length of one of longer side). For instance, the perimeter of a 3 inch by 4 inch rectangle is: (2 x 3) + (2 x 4) = 6 + 8 = 14 inches.

139. What is the perimeter of rectangle shown below? _____

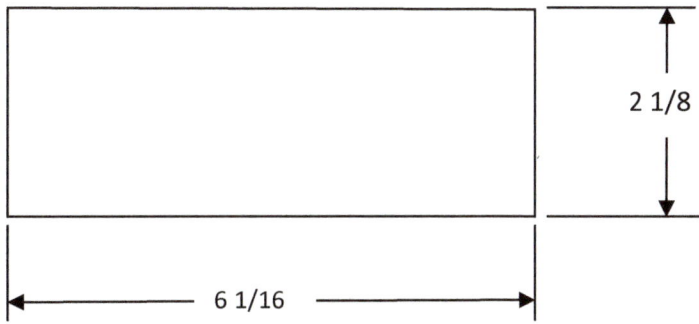

Basic Math

Multiplying and Dividing Fractions:

140. What is the combined length of the two parts below?

_____

141. How much longer is the part on the right than the part on

The left ? _____

142. If five of both parts were produced from a single length of
stock, how much material would be used ? _____

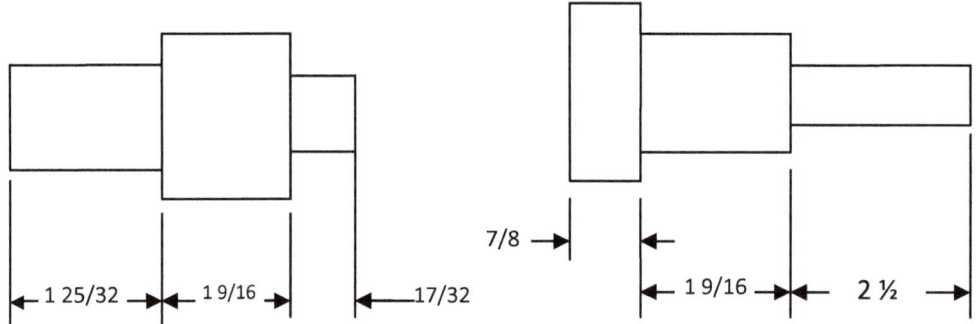

7/8 →

←— 1 25/32 —→|←— 1 9/16 —→ |←——17/32 |←— 1 9/16 —→|←— 2 ½ —→|

Basic Math

Decimal Overview:

In CNC production, only values as decimals are used. The learner must take dimensions from prints, whether fractions or whatever, and convert them to decimal values in order to insert them into the CNC control.

It is best to use "mean" dimensions (middle of tolerance range), carried out and truncated values to the machine control range of four decimal places. For instance; the dimension for 5/32 = .15625, would be entered as .1562, dropping the remaining 5 and not rounding up or down.

Definition:

Decimal – A specific form of fraction.

- Like a common fraction, a decimal expresses one or more equal parts of a whole.
- Unlike a common fraction:
  - A decimal always has a denominator of 10 or a power of 10.
  - A decimal's denominator is not written. It's shown by place values to the right of a decimal point.

The decimal $.89 is another way to express 89/100ths of a dollar. Its denominator is shown by the places where 8 and 9 appear to the right of the decimal point.

Powers of 10:

A power of 10 simply means the product of 10 multiplied by itself any number of times. 10 x 10 x 10, or 1,000, is a power of 10. So, 10 x 10 x 10 x 10 x 10, or 100,000.

Basic Math

Decimal Placement:

You're already familiar with decimals – you use them every day, whether you realize it or not. Just think of how often you work with money or dollar amounts, which are almost always expressed as decimals. Other everyday examples include:

> ➤ The distances you drive, as shown on your car's odometer in tenths of a mile.
> ➤ The amount of gasoline you put in your car, shown on the pump in tenths and hundredths of a gallon.
> ➤ The amount of energy you use, as shown on your monthly gas and electric bills.
> ➤ Interest rates that you're charged for loans or credit card charges, or that you earn on savings or checking accounts.
> ➤ The weights of many foods, such as a box of cereal, a can of soup or a package of ground beef, which may show pounds, ounces or grams as decimals.

The CNC machine shop lingo is just a little bit different with the placement values.

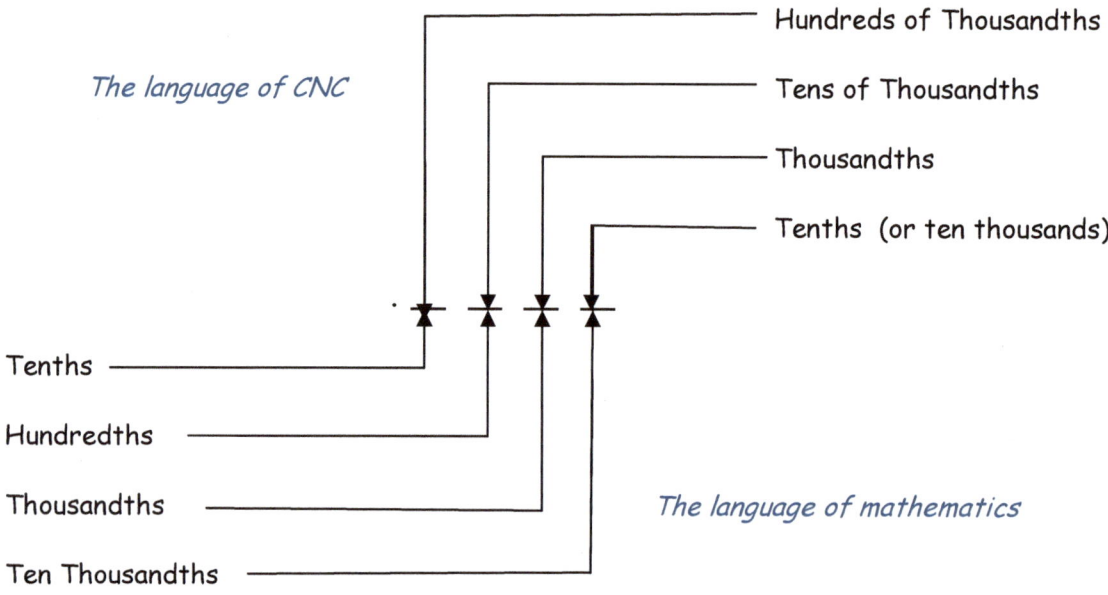

Basic Math

It's not enough to be able to work with decimals on paper. You must also know how to say the decimals correctly when you're talking about them on the CNC shop floor. Otherwise, your listener could get the wrong information.

For example, the dimension 10.3125 could be read two different ways:

> ➤ 10 inches, three hundred twelve thousandths and five tenths.
>   Or
> ➤ Ten point three one two five.

The first is more often used, but either is accepted.

Practice: (write the spoken version of these numbers as decimals)

144. 1.3125 _____

145. 5.265 _____

146. 318 _____

147. 23.9753 _____

148. 3.75 _____

149. 1/16 _____

Basic Math

Adding & Subtracting Decimals:

- Solve the following: (remember that all values are truncated to four decimal places)

150. 10.5 + 6.43 + 2.27 = _____

151. 14.536 – 8.7216 = _____

152. 1 – 0.9982 = _____

153. (19.8425 + 6.056) – 5.0123 = _____

154. You are using a 15-inch length of stock to produce four parts with overall lengths of 3.82", 3.16", 3.5" and 2.05". How much stock will you use?

155. How much stock will be left over?

156. X = _____

157. Y = _____

Basic Math

Decimal Tolerances:

A tolerance is the amount by which a part feature can vary from its specified dimension and still be considered acceptable for its production, assembly or other performance requirements. A tolerance accompanies every dimension on a part print, either stated or implied.

Types of Tolerances:
- **Limit dimensioning** – Shows the upper and lower limits for a part feature, and may be written in one of two ways.

    2.380   or   2.370 – 2.380
    2.370

- **Plus-and-minus tolerance** – Shows the specified dimension, followed by the amount of variation from the dimension that's acceptable.

    .750 +.003 in.

- **Bilateral tolerance** – A plus-and-minus tolerance in which some variation in both directions from the specified dimension is acceptable. The amount of acceptable variation may or may not be equal in both directions.

      +.003   +.0015
    .570 -.001   or   1.4720 -.0020

- **Unilateral tolerance** - A plus-and-minus tolerance in which some variation in only one direction from the specified dimension is acceptable.

    20.00 +.04 in.   48.255 -.003 in.

Whatever the style of tolerance used, we as programmers will find the middle of tolerance range (mean) and use this value to four decimal places no matter the amount of range given.

Basic Math

Decimal Tolerances:

When you work with a part that's dimensioned with mixed styles of plus-and-minus tolerances, you must add and subtract the tolerance from the specified dimension to determine the "mean" value used with cartesian coordinates in the part program.

An example of how this is achieved is shown below:

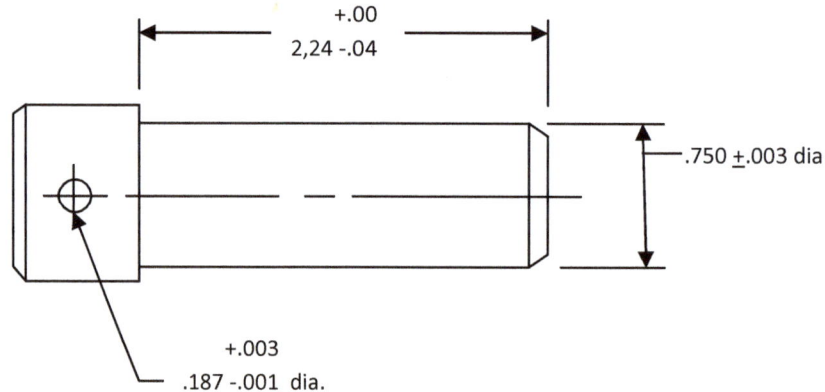

The dimension 2.24 length has an upper limit of 2.24 and a lower limit of 2.20 thus a mean value of 2.2200.

The dimension .750 has a mean diameter value of .7500 because of the equilateral tolerance.

The .187 dia hole with it's given tolerance, would probably be ok drilled with a 3/16 dia drill. A drill can usually hold size within .003 over nominal size but not produce a hole smaller than it's actual size.

Basic Math

Practice Calculating limits From Decimal Tolerances:

Dimension	Limits		Measurement
	Low	High	(mean)
161. 1.125 ± .002	_____	_____	_____
162. 2.5 +.00 −.04	_____	_____	_____
163. .060 ± .005	_____	_____	_____
164. 2.096 ± .0025	_____	_____	_____
165. .187 +.003 −.001	_____	_____	_____
166. 5.00 ± .01	_____	_____	_____

Basic Math

Using Gage Blocks:

Gage blocks are very accurate measuring devices. In production areas, they're used most often to establish a **standard** for a specified dimension, so that you can compare it against the actual measurement of a part feature.

- Supplied in sets that can range from less than 10 to more than 100 sizes.
- For most applications, several blocks are stacked together to match a desired dimension.
    - Blocks are stacked using an action called "wringing".
    - Wrung gage blocks equal their added lengths.
- **Wear blocks** may be used at one or both ends of a stack.
    - They take the wear and tear of repeated use and protect the accuracy of the rest of the blocks.
    - They're usually .050 or .100 in. long.

How to Select Gage Blocks:

Before you can stack gage blocks to make comparative measurement, you must select the sizes that will total the desired dimension, using the fewest number of blocks.

- Write down the desired dimension.
- If the stack will include wear blocks at one or both ends, subtract their length from the desired dimension first.
- From right to left, select a block that will reduce to zero the last one or two digits of the number.
- Subtract the length of the selected block from the dimension.
- Again, select a block that will reduce to zero the last one or two digits of the number, then subtract its length from the number.
- Continue selecting blocks in the same manner, working from the right to left digits of the remaining portion of the dimension.
- When all the blocks are selected, add their lengths to make sure they total the desired dimension.

Basic Math

Basic Math

Using Gage Blocks:

Problem:

You need to create a stack of gage blocks to match a specified dimension of 2.7356 in., from the 81 – piece set shown below. The stack will include a .100 in. wear block on one end. What combination of sizes would you choose?

81 – Piece Gage Block Set

.1001 .1002 .1003 .1004 .1005 .1006 .1007 .1008 .1009

,100 .101 .102 .103 .104 .105 .106 .107 .108 .109

.110 .111 .112 .113 .114 .115 .116 .117 .118 .119

.120 .121 .122 .123 .124 .125 .126 .127 .128 .129

.130 .131 .132 .133 .134 .135 .136 .137 .138 .139

.140 .141 .142 .143 .144 .145 .146 .147 .148 .149

.050 .100 .150 .200 .250 .300 .350 .400

.450 .500 .550 .600 .650 .700 .750 .800

,850 .900 .950

1.000 2.000 3.000 4.000

Solution:

2.7356	Write the desired dimension.
- .100	Subtract the length of the wear block first.
2.6356	Look for a block that reduces the last digit or two to zero (right to left).
- .1006	Subtract its length.
2.5350	Again, look for a block that reduces the last digit or two to zero.
- .135	Subtract its length.
2.4000	Again, look for a block that reduces the last digit or two to zero.
- .400	Subtract its length.
2.0000	Again, look for a block that reduces the last digit or two to zero.
-2.0000	Subtract its length.
0.0000	Check by adding the lengths selected or by measure.

Basic Math

Using Gage Blocks:

Practice: Reduce the number from moving from right to left (use the least amount of blocks)

2.855   1.7662

3.4118   5.6525

Basic Math

Using Gage Blocks:

Dimension	limits		Measured	Acceptable
	Low	High	Value	(yes or no)
167. 2.025 $\pm$ .0035	_____	_____	2.023	_____
168. 5.25 -.04	_____	_____	5.197	_____
169. .862 $\pm$.002	_____	_____	.863	_____
170. 3.016 $\pm$ .003	_____	_____	3.015	_____
171. 12.175 +.0025	_____	_____	12.173	_____
172. 9.625 $\pm$ .0015	_____	_____	9.624	_____
173. 5.125 + .002	_____	_____	5.124	_____

Basic Math

Using Gage Blocks:

Sizes in a 36-Piece Set of Gage Blocks

.0501 .0502 .0504 .0506 .0507 .0508 .0509

.0510 .0520 .0530 .0540 .0550 .0560 .0570 .0580 .0590

.0500 .0600 .0700 .0800 .0900

.1000 .1100 .1200 .1300 .1400 .1500

.2000 .3000 .4000 .5000

1.0000 2.0000 3.0000

174. Using the sizes in the 36-piece set of gage blocks shown above, select an appropriate combination for the following dimensions.

1.6055   2.7214

Basic Math

Dividing Decimals:

Problem:

You're using X-bar/R charts to monitor a process, and you need to find an X-bar, or average, value for the following five readings that make up the subgroup:

2.86   2.78   2.81   2.84   2.78

Solution:

To find an average, add the values, then divide by the number of values.

The sum of the measurement readings above is the decimal 14.07.

It's divided by 5, the number of values, to find the subgroup average of 2.814.

Problem:

You have a length of stock that measures 55.125 in. With it, you will be producing parts that each have an overall length of 3.15 in. How many parts can be made from the length of stock?

Solution:

$$55.125 \div 3.15 = 17.5 \text{ or } 17 \text{ parts}$$

Basic Math

Multiplying Decimals:

175. In each of the multiplication problems below, write the decimal point in its proper place in the answer. If necessary, add zeros to fill decimal places.

96.81 x .035 =   338835   340 x .002 =   0680

.068 x .075 =   5100   191.7 x .0005 =   9585

176. If a food processing operation uses an average 0f 12.45 lbs. of sugar every 20 minutes, how many pounds of sugar are used during a 40-hour period?

177. How many pieces of wire, each 5.15 in. long, can be cut from a coil that's 252.35 in. long?

178. The formula for finding the circumference of a circle, or the distance around its edge, is to multiply its diameter by the number 3.1416 (approximately). What is the circumference of a shaft with a diameter of .775 in.?

Basic Math

More Exercises Using Decimals:

179. You're in charge of planning the food for a lunchtime meeting to celebrate the accomplishments of an 18-person work group. How much will it cost for all the items listed below?
    o   2 deli trays, $18.50 each
    o   3 pizzas, $16.95 each
    o   2 large salads, $5.75 each
    o   6 six-packs of canned beverages, $2.29
    o   Paper goods, such as napkins and plates, totaling $8.76

180. How much will the food, beverages and paper goods cost per person, rounding off to the nearest penny?

181. A drill has a diameter of .3575 in. when measured with a vernier micrometer. What is its equivalent fractional size, to the nearest 1/64 in.?

182. A single can of motor oil costs $1.19, and a case costs $11.49. How much money would you save on each can if you bought the case instead of 12 single cans?

183. What's the total amount you would save by buying a case instead of 12 single cans?

Basic Math

More Exercises Using Decimals:

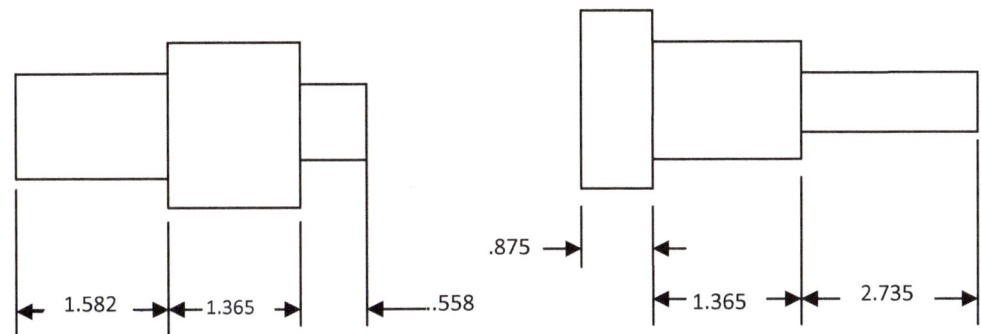

184. What is the combined length of these two parts? _____

185. How much longer is the part on the right than the part on the left? _____

186. If a 54 in. length of stock was used to produce five of each of the two parts, how much stock would be left over? _____

187. If the dimension of 2.6645 in. has a tolerance of +.0025/-.0015, what are its high and low limits?

        High limit = _____   Low limit = _____

The

End

www.ingramcontent.com/pod-product-compliance
Lightning Source LLC
Chambersburg PA
CBHW051059180526

45172CB00002B/698